NATIONAL GEOGRAPHIC

Ladders

LET'S KEEP MOVING

GENRE **Reference Article**

Read to find out about some cars that represented great technology and style in the past.

Cool Cars

by Jennifer Boudart

Inventors began creating automobiles, or cars, in the late 1800s. At that time, they were little more than "horseless carriages." But big changes would come in the future. Let's look at some of the features of cars in the 1940s, 50s, and 60s that made them safe, affordable, and, well, cool!

The company said that the car was named for the strength of the pilgrims at Plymouth Rock. Like the 1942 model, the hood ornament sported a sailing ship—the Mayflower. But now the hood ornament was clear-plastic and lit up.

Farmers used a kind of string called Plymouth twine to bind hay. Legend says the car was given the same well-known name so farmers would buy it.

1947

Plymouth DeLuxe Sedan

When World War II ended, automakers returned to manufacturing cars after a four-year pause. The public was eager to buy new models. Yet automakers had neither the time nor the money to invest in new car designs. Chrysler Corporation scrambled to re-tool its factories, which had been turning out military equipment such as airplanes and tanks during the war. They updated their 1942 models with a few new features. New grilles, different headlights, and new hood ornaments made buyers think they were getting a different car.

FEATURES
Seating for six
Interior mirror adjusters
Raised bench front seat so driver rides high
Pushbutton AM radio searches for stations
Price: Around $1,200

The gas cap covered the fuel tank, which held about 53 liters (14 gallons) of gas.

The car featured brakes designed to respond to lighter foot pressure and slow the car more quickly than brake systems on other cars.

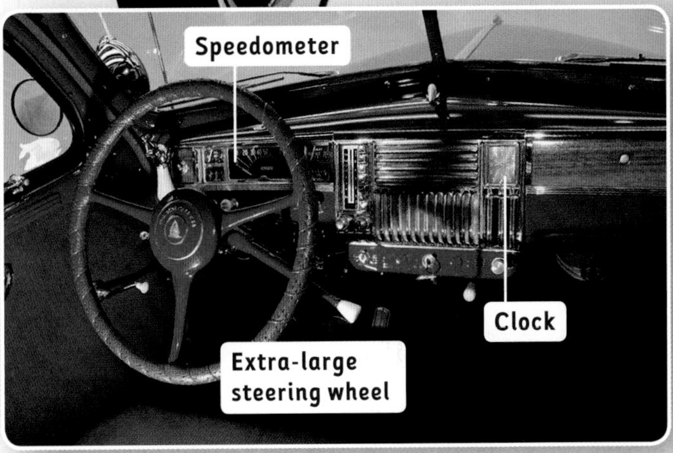

1956
Cadillac Eldorado Seville

For anyone looking to buy a luxury car in the fifties, one name came to mind: Cadillac. This car company specialized in producing flashy cars. Its cars had lots of chrome, eye-catching lines, and powerful engines. Cadillac established itself as a leader in luxury car sales during this decade. In 1956, Cadillac released a two-door hardtop model called the Eldorado Seville. The Eldorado stood out from the pack. It had extensive chrome trim, stylish tail fins, and an extra-powerful engine. Its popularity helped Cadillac earn record sales in 1956. These sales bumped the automaker up from 10th to 9th place in sales of American-made cars.

FEATURES
Six-way power seats
Electrically powered radio antenna
Seating for six
Whitewall tires
Pushbutton AM radio that searches for stations
Seatbelts (brand new option)
Air conditioning
Price: Around $6,500

The Eldorado's tail fins were longer and taller than any other model released in 1956.

The gas cap was hidden in a taillight.

Cadillac reached for daring new heights with the Eldorado's tail fins. They looked a bit like shark fins, and would become even bigger in later models. They actually improved handling by making the car more **aerodynamic.** The aerodynamic shape allowed air to flow around the car so the car was easier to steer and used less gas. The fins also helped the car to be steadier at higher speeds.

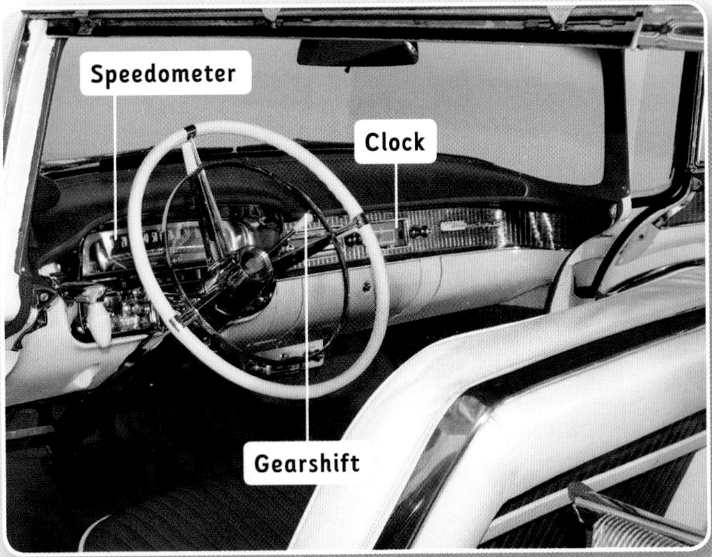

Treated safety glass gave windows a green tint that filtered sunlight, reducing glare and heat. The result was a safer, cooler ride.

An automatic headlight dimmer was mounted on the dashboard. It detected light from oncoming cars. The incoming light triggered an **electric current** that turned on a dimming switch. The Eldorado's lights automatically dimmed until the other car passed, then brightened again. That way, bright headlights didn't bother oncoming drivers.

Buyers could choose a gold-plated grille. Bullet-shaped bumper guards offered protection, plus a sense of style and speed.

The Eldorado's bold, spoked wheels came with silver or gold plating.

In Spanish, the words "el dorado" mean "the golden one."

1966

FORD MUSTANG CONVERTIBLE

Introduced just two years earlier and still being sold today, more people bought a Mustang in 1966 than in any other year. The Mustang appealed to the younger generation of the day because it was so different from earlier cars. Mustangs were just as exotic as foreign sports cars, but they also had a back seat and plenty of design options. Ford offered three models—a coupe, a fastback, and a convertible. Mustang owners felt they were creating a custom car as they picked features such as paint color, wheel style, and engine size.

FEATURES
Front seatbelts
Floor completely carpeted
Electrically powered convertible top
Seating for four, including bucket (one-person) seats in front
Pushbutton AM radio that searches for stations
Price: Around $2,650

The 1966 grille showed the running mustang figure. Horizontal chrome bars gave the car a sleek look.

The wheels' racecar-style covers pleased buyers.

The Mustang has been produced every year since it was introduced in 1964.

The Mustang's design and logo inspired a new car style called "pony cars," with long hoods and short back ends. They were fast, too. Depending on engine type, Mustangs hit speeds up to 135 miles per hour.

On March 2, 1966, the one-millionth Mustang rolled off the assembly line.

In addition to Candy Apple Red, people could order one of 22 other colors. They included Raven Black, Tahoe Turquoise, and Springtime Yellow.

Check In How were these three classic cars similar and different?

GENRE Third-Person Narrative

Read to find out how a young man made his dream come true.

FLIGHT AROUND THE GLOBE

by Judy Elgin Jensen

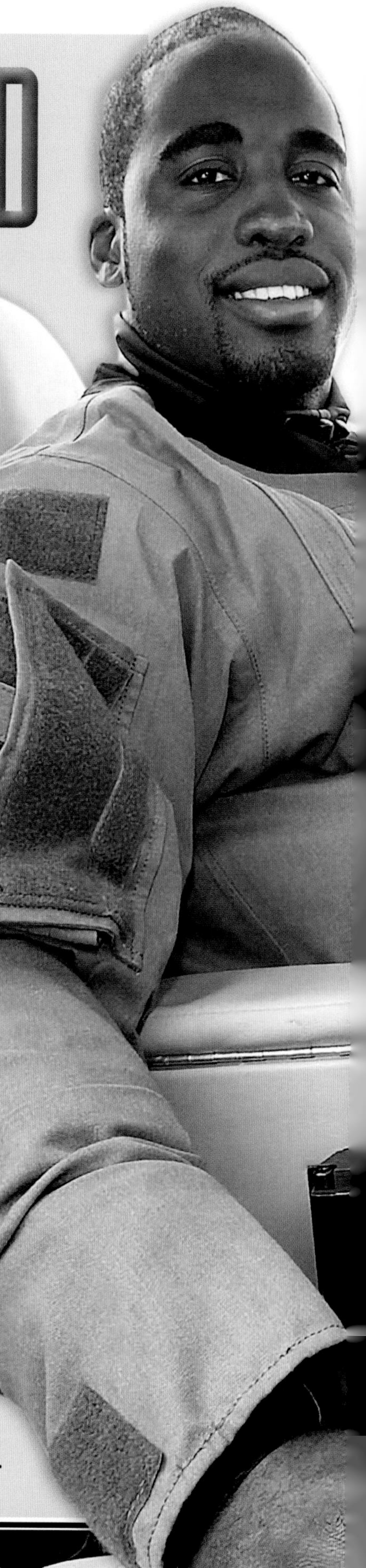

On a spring day in 2007, Barrington Irving opened the cockpit door of an airplane named *Inspiration.* He was about to fly into history. Perhaps Irving recalled another door that day, the one to his parents' bookstore. Eight years earlier, a pilot had walked through that door. At the time, 15-year-old Irving dreamed of playing college football. The pilot suggested a career in **aviation** to Irving, who replied that he wasn't smart enough. So the pilot invited Irving aboard an airplane, and the young man was hooked.

Instead of taking a football scholarship, Irving worked several jobs to earn money. He spent time at the local airport and even played video games to practice his flight skills. Then Irving earned a college scholarship to study aviation science and take flying lessons. During his senior year, 20-year-old Irving's new dream took shape.

BARRINGTON IRVING is a pilot and an educator. In 2007, at age 23, Irving became the youngest person and first African American to fly solo around the globe. Irving also founded "Experience Aviation," a nonprofit organization that offers programs in aviation for middle school and high school students.

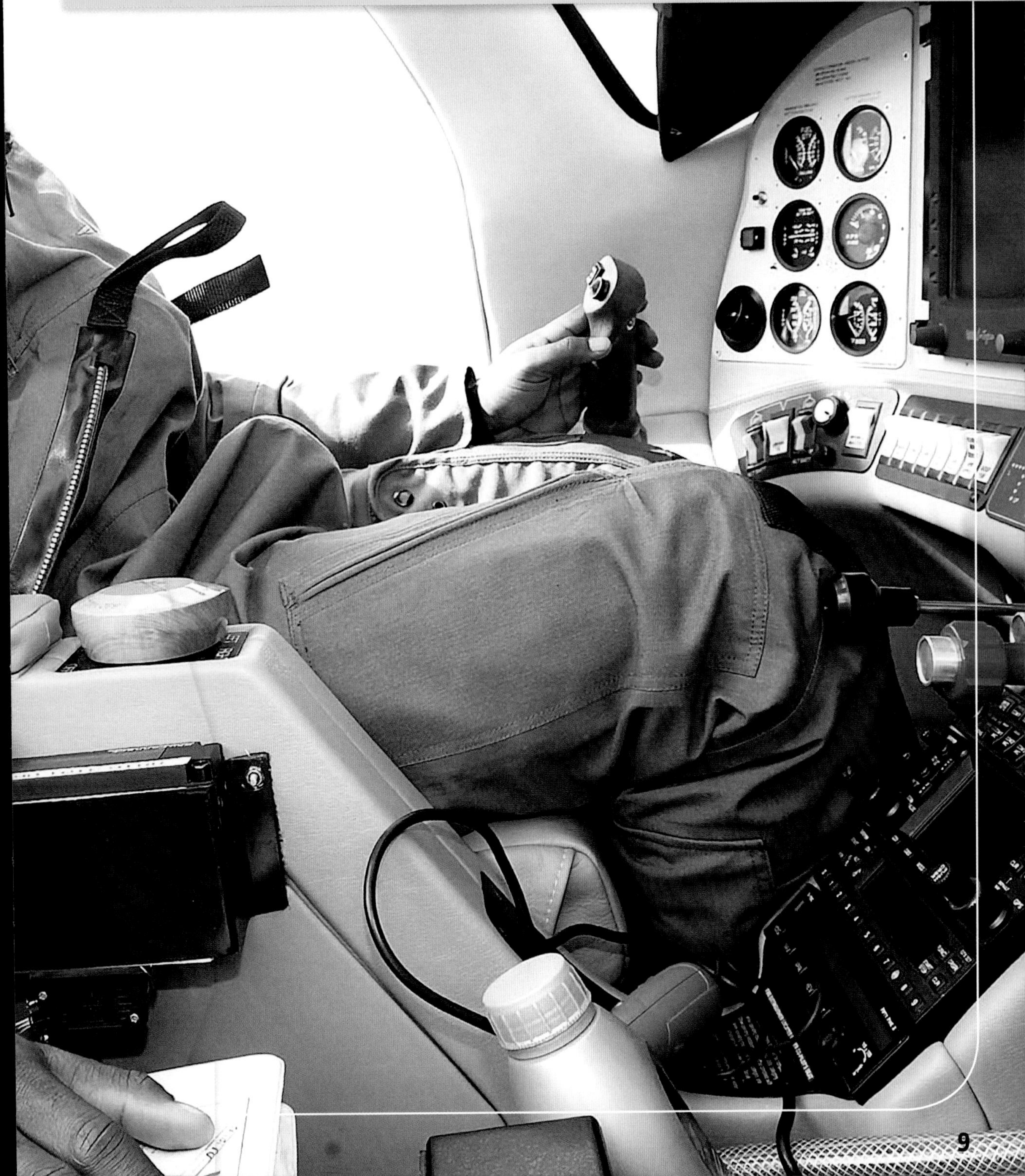

Irving wanted to become the youngest person to fly solo around the world. His dream seemed impossible. For one thing, Irving was incredibly young for such a feat. No one younger than the mid-thirties had ever done it. For another thing, Irving had no money and no airplane!

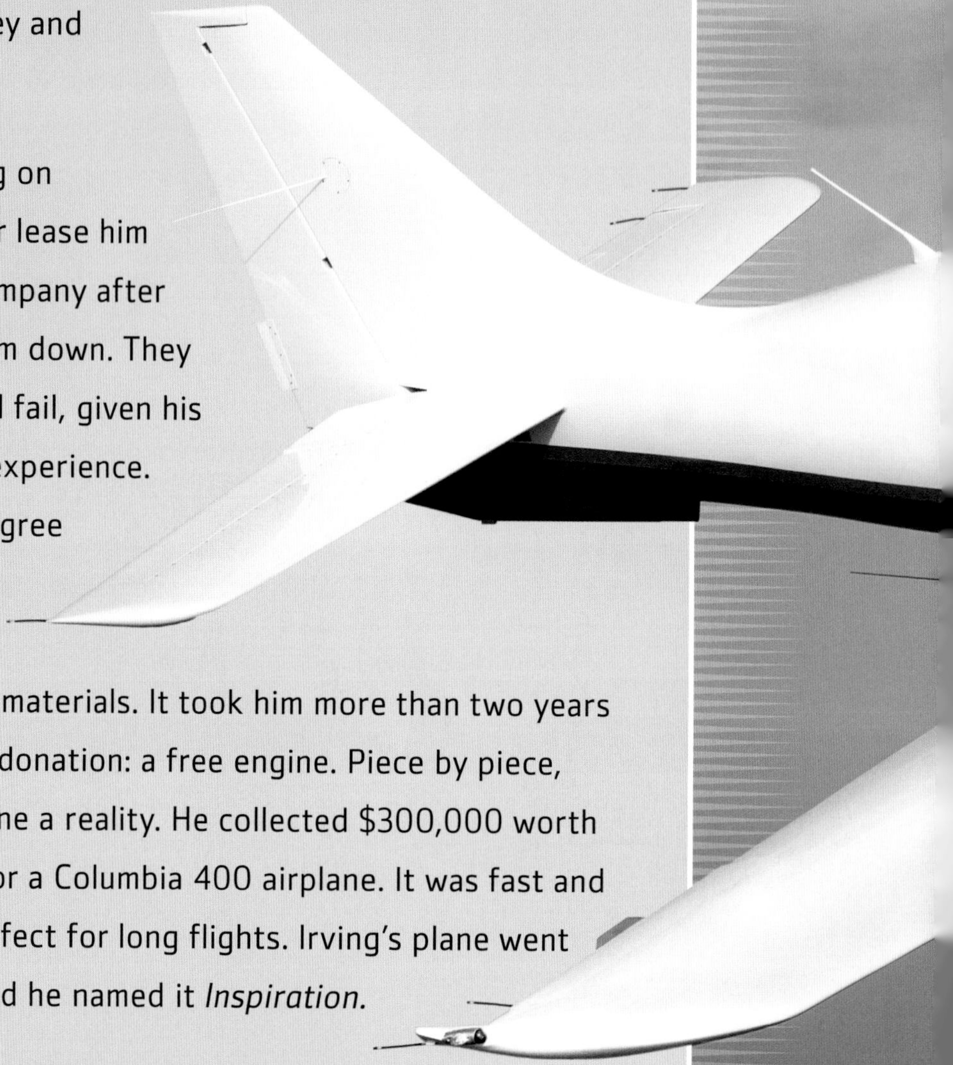

Irving was counting on sponsors to lend or lease him an airplane, but company after company turned him down. They were sure he would fail, given his youth and limited experience. One company did agree to build a plane for free, but Irving had to provide the materials. It took him more than two years to receive his first donation: a free engine. Piece by piece, Irving made his plane a reality. He collected $300,000 worth of donated parts for a Columbia 400 airplane. It was fast and fuel-efficient—perfect for long flights. Irving's plane went into production, and he named it *Inspiration.*

While *Inspiration* was being built, Irving learned all he could about its operation. Most pilots attempting such a trip would have had 10,000-plus flight hours; Irving ended up having 600. He practiced flying over mountains, oceans, and in different weather conditions. He spent a year arranging stops in many countries, where he'd rest, refuel, and meet local people. Finally, on March 23, 2007, thousands of spectators watched 23-year-old Irving take off from Miami, Florida.

This airplane is similar to Irving's plane, *Inspiration*.

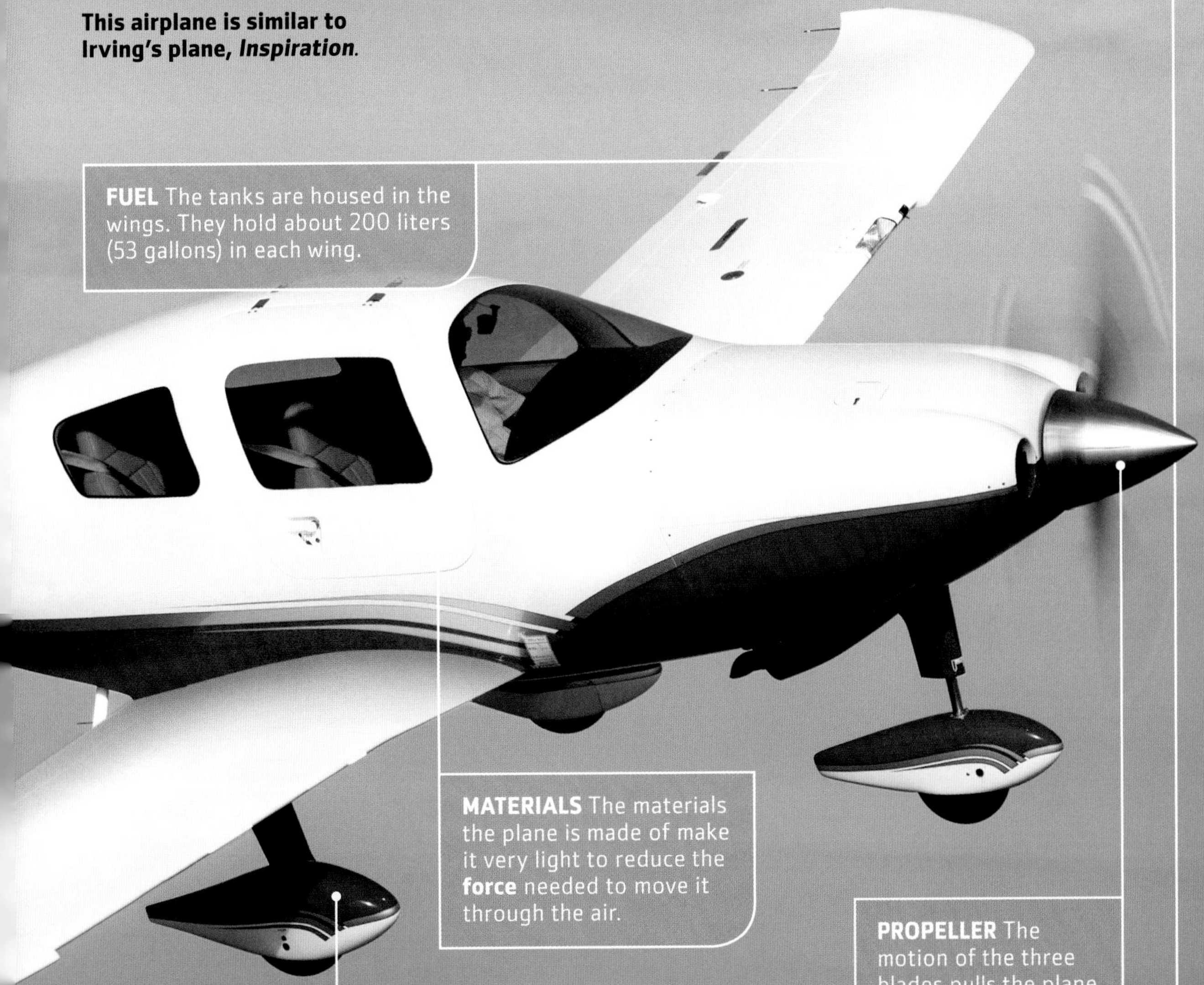

FUEL The tanks are housed in the wings. They hold about 200 liters (53 gallons) in each wing.

MATERIALS The materials the plane is made of make it very light to reduce the **force** needed to move it through the air.

PROPELLER The motion of the three blades pulls the plane through the air.

SHAPE The covers over the wheels are one of the features that make the plane more **aerodynamic,** which helps it fly faster on less fuel than other similar planes.

Inside the Cockpit

GAUGES Energy stored in a battery creates an **electric current** that runs lights, flight instruments, and a radio. The gauges help the pilot know if all the systems are working correctly.

Irving hoped to return to Miami in 41 days, but challenges arose quickly. Heavy snows forced a two-week delay in Canada—just his third stop. From there, Irving flew about 2,250 kilometers (1,400 miles) to the Azores—nine straight hours over the ocean. He made the plane climb high to avoid storms because weather was a big concern. *Inspiration* lacked weather radar and de-icing systems because no one had donated this equipment and Irving couldn't afford it. He would in fact face plenty of bad weather during his trip. While flying over Saudi Arabia, he hit a sandstorm. Had sand clogged the engine, he could have crashed!

Irving battled loneliness and fatigue from sitting in a cramped cockpit for up to 12 hours and flying on as little as four hours of sleep. At each stop, Irving updated his blog, where 300,000 students tracking *Inspiration* online could see pictures and video. He also chatted with fans, met local officials, and prepped for the next flight. On his longest leg—a 12-hour flight from Japan to Alaska—powerful winds rocked the plane and storms limited visibility, but Irving arrived safely. A few weeks later, on June 27, Irving made his final landing, back in Miami. He'd been gone 97 days.

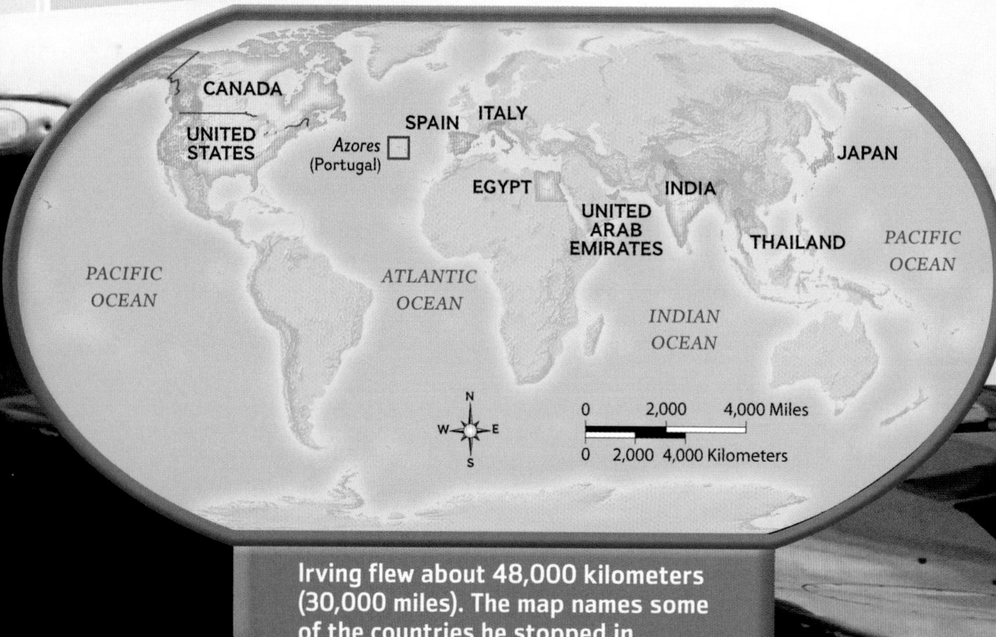

Irving flew about 48,000 kilometers (30,000 miles). The map names some of the countries he stopped in.

Inspiring Others

Irving was inspired by the many young people he met while planning his trip. In 2005, he founded Experience Aviation, a nonprofit organization that teaches young people about careers in math and science. Middle and high school students attend camps where they can build airplanes, take field trips, and meet aviation professionals. What's next? Irving is converting a jet into a flying classroom. Students will follow online as Irving flies around the world once again.

Check In What challenges did Irving overcome to make his dream come true?

GENRE History Article

Read to find out about the history of NASA's shuttle program.

The Space Shutt

1981–2011

by Jennifer Boudart

It's a rocket! It's a plane! It's a giant space taxi! Actually, the space shuttle was all three. This reusable spacecraft, built by the National Aeronautics and Space Administration (NASA), launched like a rocket, landed like an airplane, and moved people and materials between Earth and the International Space Station (ISS).

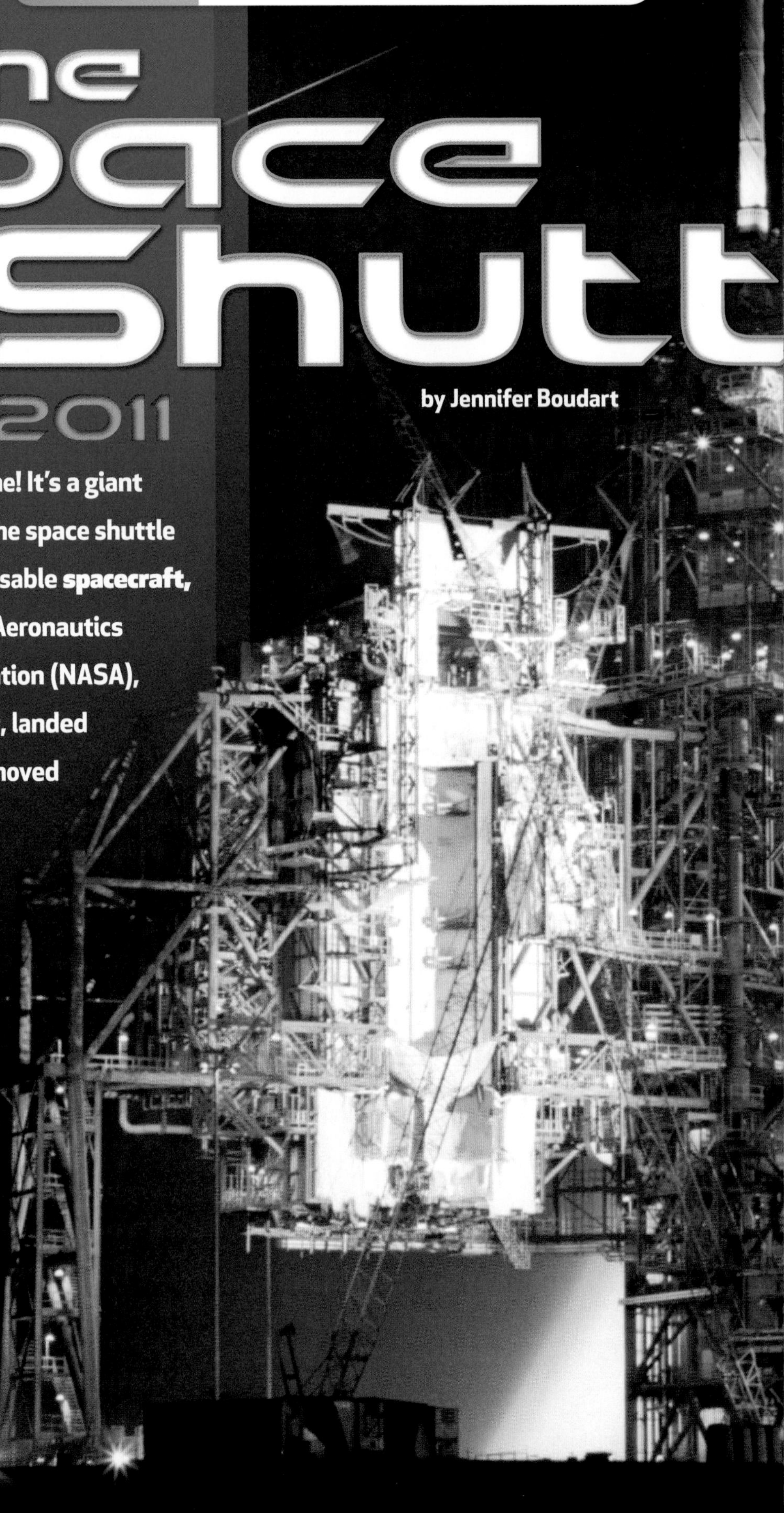

Columbia sits on Launch Pad 39A in March 1981. Everything is being readied for its first flight.

From 1981 to 2011, NASA sent five different shuttles on 134 missions. Hundreds of people orbited Earth this way. They spent days and even weeks running investigations, studying Earth, launching space equipment, and even helping to build a space station. Let's meet the shuttles!

ENTERPRISE (1976–1979)
Full-sized test vehicle used for flight tests in the atmosphere and on the ground.

COLUMBIA (1979–2003)
Missions: 28
Distance Flown: 195,852,325 km (121,696,993 mi.)
Total Passengers: 159

CHALLENGER (1982–1986)
Missions: 10
Distance Flown: 38,079,155 km (23,661,290 mi.)
Total Passengers: 60

DISCOVERY (1983–2011)
Missions: 39
Distance Flown: 238,539,663 km (148,221,675 mi.)
Total Passengers: 222

ATLANTIS (1985–2011)
Missions: 32
Distance Flown: 194,168,813 km (120,650,907 mi.)
Total Passengers: 191

ENDEAVOUR (1991–2011)
Missions: 25
Distance Flown: 197,761,262 km (122,883,151 mi.)
Total Passengers: 145

Mission Milestones

September 17 1976

Enterprise rolls off the assembly line. Never equipped to fly in space, it later supplied some parts for shuttles that did.

February 7 1984

Two astronauts aboard *Challenger* took the first untethered spacewalk. Bruce McCandless II and Robert L. Stewart floated freely, using hand-controlled jet packs on their spacesuits to steer themselves for almost six hours!

April 12 1981

Columbia blasted off on NASA's first space shuttle mission with two astronauts aboard. The shuttle flew just over 1.6 million kilometers (1 million miles) in two days and orbited Earth 37 times.

January 28 1986

Challenger exploded shortly after launching. The shuttle and its seven-person crew were lost.

October 18 1989

From space, *Atlantis* launched the first spacecraft to Jupiter. Named *Galileo*, the unmanned spacecraft shot off toward Jupiter, entering that planet's orbit six years later.

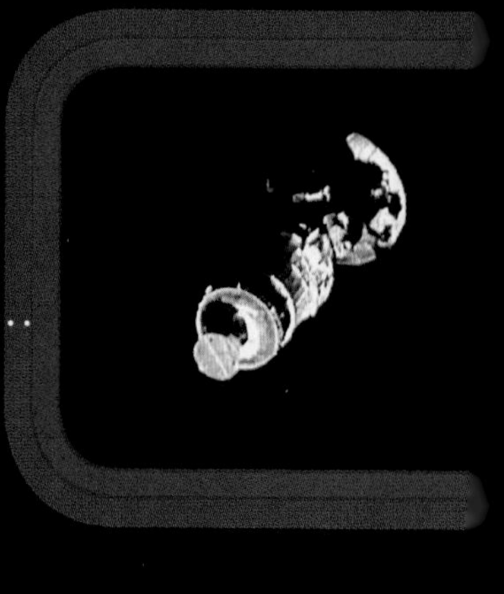

April 25 1990

Discovery launched the Hubble Space Telescope. Since that launch, the telescope has spied on black holes, far-off galaxies, dying stars, and planets beyond our solar system. Space shuttles have visited the telescope a number of times to make repairs and install updated parts.

October 29 1998

Aboard *Discovery*, John Glenn became the oldest person in space at age 77. In 1962, he was the first U.S. astronaut to orbit Earth.

February 11 2000

An international crew from the United States, Europe, and Japan blasted off aboard *Endeavour*. They spent 11 days creating a three-dimensional map of Earth's surface, using two radar antennas attached to the shuttle. The radar mapped about 80 percent of Earth's surface, such as this river and mountain range in Russia. It took four years to analyze all the pictures!

February 1 2003

Columbia broke apart while returning to Earth. The shuttle and seven-person crew were lost.

July 21 2011

Atlantis launched a satellite that would test new solar technology. Then, the shuttle glided back to Earth, ending the shuttle era.

Building a New Place in Space

The International Space Station (ISS) is the largest spacecraft ever built. It's the size of a football field and weighs almost 450,000 kilograms (1,000,000 pounds). It's made up of linking pieces called modules and it took 12 years to build. Space agencies representing 15 countries did the work.

Russia sent the first module for the ISS into space in 1998. A few weeks later, *Endeavour* brought a second module, which was connected to the first one. Then rockets and other shuttles traveled to the ISS. From 1998 to 2011, *Discovery*, *Endeavour*, and *Atlantis* made 37 trips. They transported supplies, parts for assembly, and ISS crew members.

Astronauts make repairs to the ISS during this February 2011 mission.

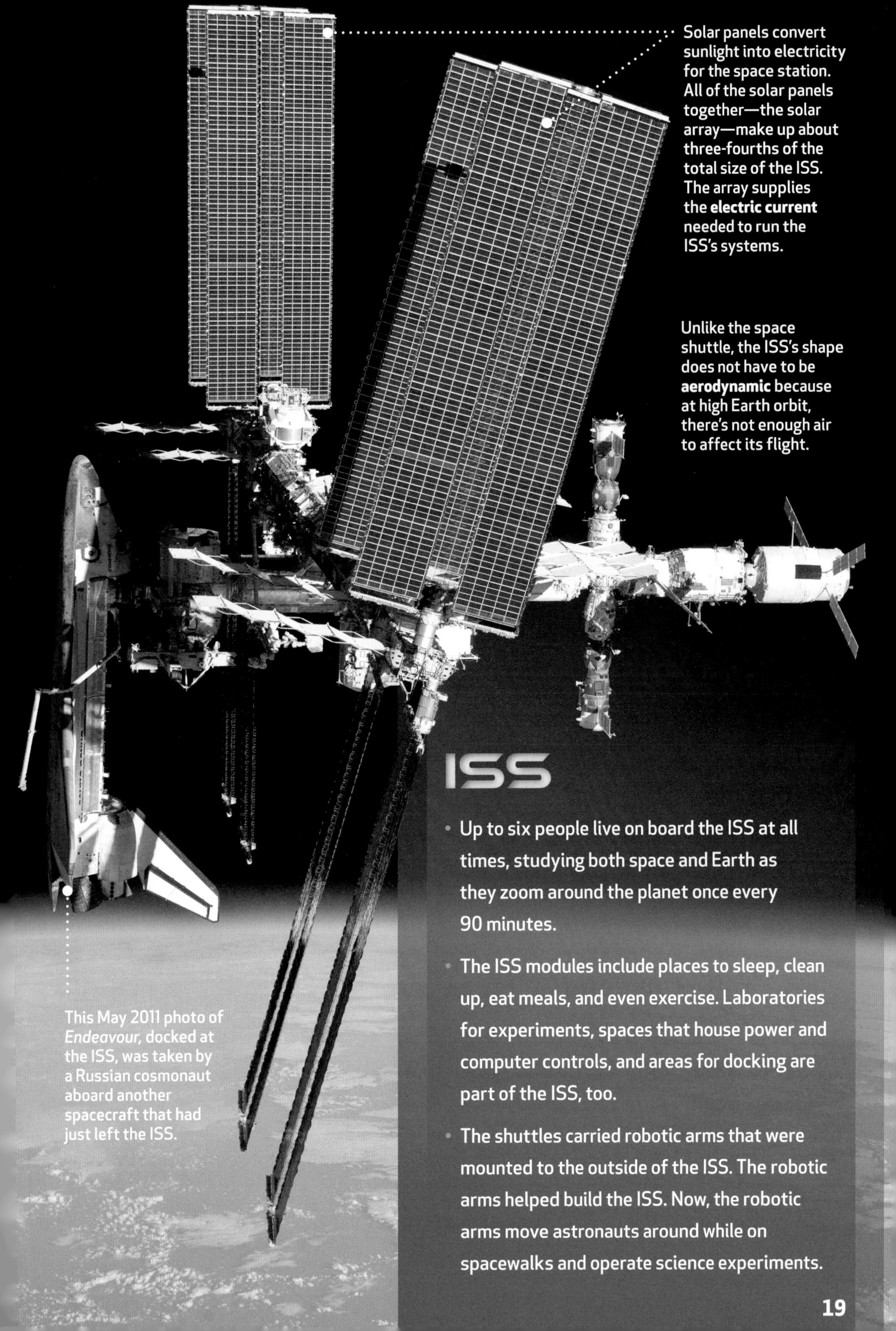

Solar panels convert sunlight into electricity for the space station. All of the solar panels together—the solar array—make up about three-fourths of the total size of the ISS. The array supplies the **electric current** needed to run the ISS's systems.

Unlike the space shuttle, the ISS's shape does not have to be **aerodynamic** because at high Earth orbit, there's not enough air to affect its flight.

This May 2011 photo of *Endeavour*, docked at the ISS, was taken by a Russian cosmonaut aboard another spacecraft that had just left the ISS.

ISS

- Up to six people live on board the ISS at all times, studying both space and Earth as they zoom around the planet once every 90 minutes.
- The ISS modules include places to sleep, clean up, eat meals, and even exercise. Laboratories for experiments, spaces that house power and computer controls, and areas for docking are part of the ISS, too.
- The shuttles carried robotic arms that were mounted to the outside of the ISS. The robotic arms helped build the ISS. Now, the robotic arms move astronauts around while on spacewalks and operate science experiments.

Mission Accomplished

On July 8, 2011, *Atlantis* fired up its boosters and left the launch pad at NASA's Kennedy Space Center. So began mission number 135. A crew of four rode on board, along with supplies and parts for the ISS. The work was routine, but the mission was not. It was, in fact, NASA's last space shuttle flight. After 30 years of working in Earth's orbit, NASA was ready to focus on sending astronauts further into space—further than the space shuttles could take them. The ISS was fully operational now. Other spacecraft from around the world would supply it as needed.

Atlantis headed for the ISS and safely docked there two days later. For eight days, crews from both spacecrafts worked to transfer the huge payload from *Atlantis* to the ISS. President Barack Obama phoned to congratulate everyone on board. *Atlantis* was loaded for its return trip to Earth's surface before separating from the ISS a final time. After launching one last satellite, *Atlantis* touched down on July 21, 2011. *Atlantis,* and the rest of the space shuttles, were home to stay.

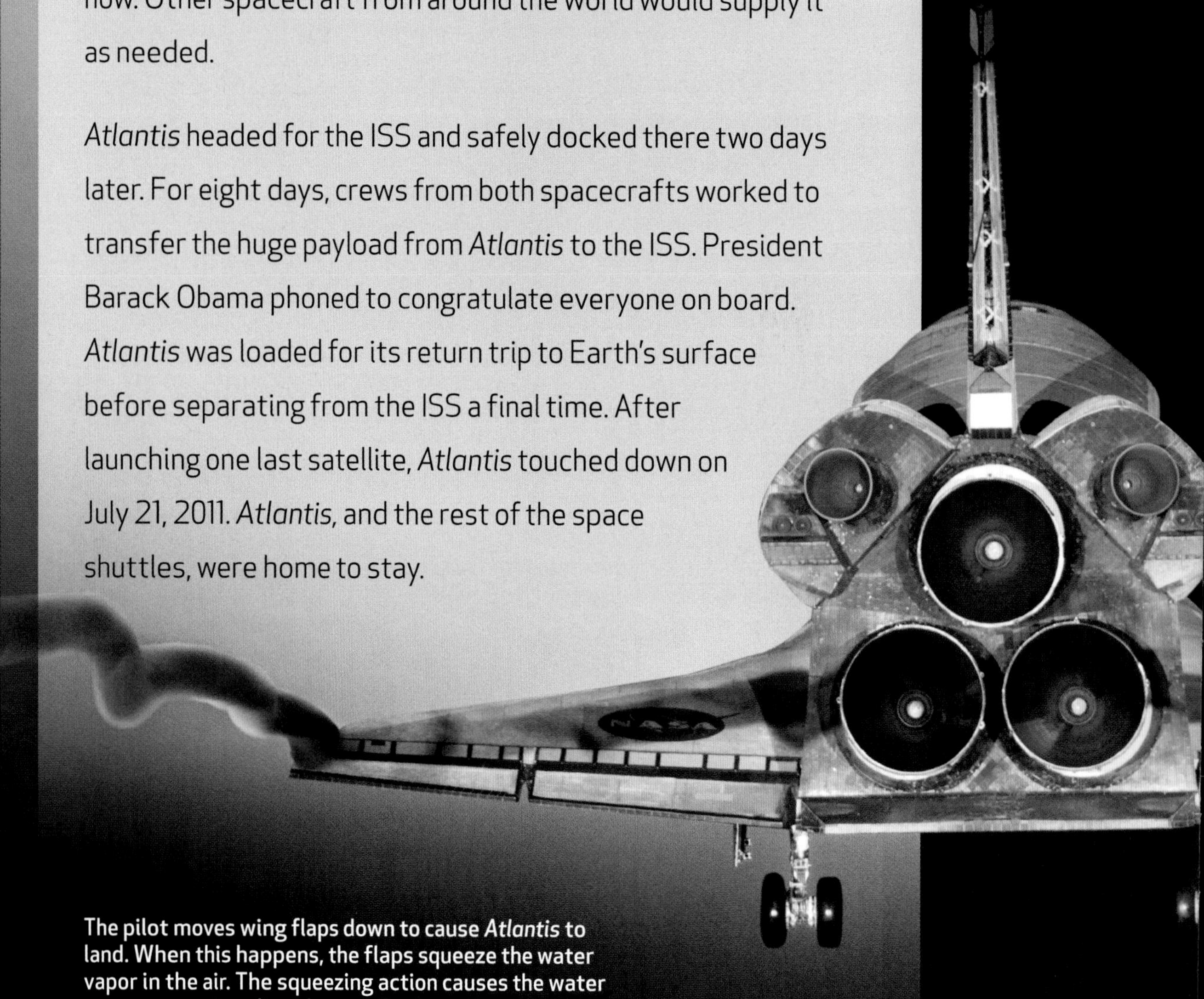

The pilot moves wing flaps down to cause *Atlantis* to land. When this happens, the flaps squeeze the water vapor in the air. The squeezing action causes the water vapor to turn into liquid water. So the curly trails from the wings are like clouds.

^ *Atlantis'* home is at the Kennedy Space Center, where she can be toured at the Visitor Complex.

A New Mission

The shuttles no longer fly, but they're still working at their final homes as tools for learning about space. Visit!

- **ATLANTIS** Kennedy Space Center Visitor's Complex (Kennedy Space Center, Florida)
- **DISCOVERY** National Air and Space Museum, Steven F. Udvar-Hazy Center (Chantilly, Virginia)
- **ENDEAVOUR** California Science Center (Los Angeles, California)
- **ENTERPRISE** Intrepid Sea, Air & Space Museum Complex (New York City, New York)

Check In How did space shuttles help space exploration?

GENRE Engineering Article | **Read to find out** how scientists solved problems to explore Mars.

NEXT STOP MA

Martian soil is loaded with red iron oxide. This is why Mars is nicknamed the red planet. Violent winds often whip up giant clouds of red dust that surround the planet.

by Judy Elgin Jensen

RS

Is life possible on Mars?

Maybe. People have dreamed about it and scientists are trying to find out. Mars is our closest planetary neighbor. In some ways it's like our own planet. Mars is half the size of Earth, but both planets have about the same amount of dry land. Their atmospheres are made up of similar gases, too, though Martian air contains deadly amounts of carbon dioxide. Both planets experience seasons and have a similar day length. Mars has volcanoes, canyons, mountains, and ice caps like those on Earth. Of course, Earth's environment supports life. What about Mars? Scientists think Mars may once have had liquid water. Water may still remain deep underground. If so, microscopic life may have existed there in the past—or may even exist today. The National Aeronautic and Space Administration (NASA) is exploring Mars for evidence of life. Its mobile science lab, the *Curiosity* Rover, has been trekking across Mars since August 5, 2012.

THE LONG ROAD TO LIFTOFF

Curiosity is not the first rover to land on Mars, but it's definitely the most high-tech. It's bigger and more mobile than the others and it carries a record number of scientific instruments. NASA spent nearly 10 years preparing for *Curiosity*'s mission.

Engineers had to build *Curiosity* and a **spacecraft** to carry it. Scientists had to calculate a landing date when Earth and Mars would be aligned—something that happens only every 26 months! Scientists also had to chart a course through space. The trip would take 254 days and cover 566,500,000 kilometers (352,000,000 miles). And there was also the question of where and how to set *Curiosity* down. NASA had never landed a rover as big as a car before. Expectations and emotions were running high on launch day, November 26, 2011.

READY...

Technicians work on Curiosity in the cleanroom at the Jet Propulsion Laboratory in California. They must avoid contaminating the rover with germs so the germs aren't carried to Mars.

SET...

Workers pack Curiosity and its spacecraft inside a protective cone before placing it atop a rocket.

A Kansas 6th-grader won an essay contest. She named *Curiosity* and signed the rover!

GO!

A rocket carrying *Curiosity* lifts off from Florida's Cape Canaveral Air Force Station on November 26, 2011. After about 45 minutes, the spacecraft carrying *Curiosity* breaks away.

THE JOURNEY

Curiosity was officially on its way! For nearly nine months, scientists kept the spacecraft on track and monitored its systems. Finally, the spacecraft reached Mars' outer atmosphere. It was time to start the entry, descent, and landing (EDL) phase. Previous rovers had been dropped to Mars, with airbags to cushion their fall. But *Curiosity* was too big and heavy to drop. NASA was trying a new method. Because the mission costs were so great, NASA was really scared it wouldn't work!

First, **friction** slowed *Curiosity* as it entered Mars' atmosphere. Then, a giant parachute opened at the back of the spacecraft, slowing it even more. Next, the protective heat shield popped off. *Curiosity* was still tucked inside the back half of the spacecraft. It was connected to a sky crane, a piece of equipment with rockets to control its own speed and direction. The sky crane ejected next and fired up its rockets. About six stories above the surface, the crane lowered *Curiosity* on a set of cables. The rover's wheels touched down, the cables came away, and the sky crane flew off. *Curiosity* landed right on target, in an area called Gale Crater.

NASA could only wait to see if their landing method worked. They called this time the "7 Minutes of Terror."

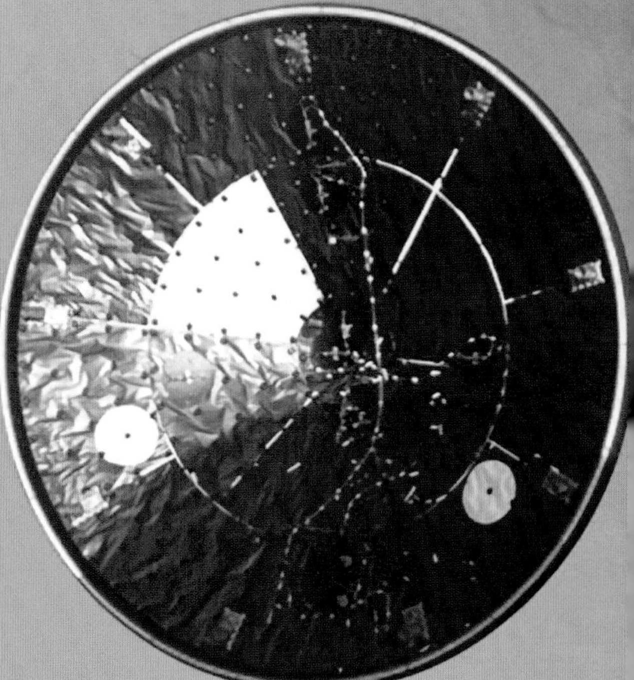

A camera on *Curiosity* took this photo of the heat shield dropping away. The 7 Minutes of Terror was about over!

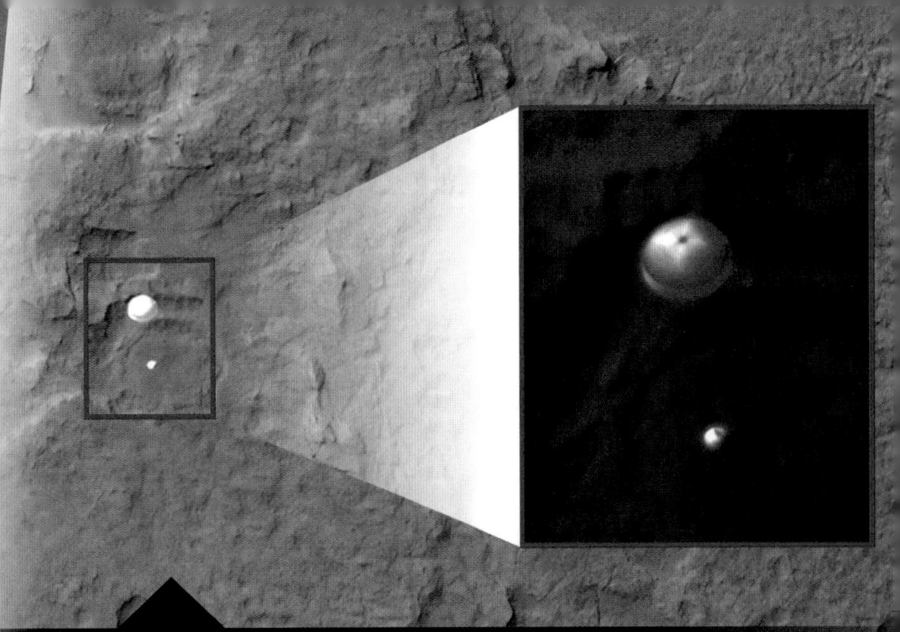

A satellite orbiting Mars took this photo of the parachute slowing down *Curiosity's* descent.

The spacecraft entered Mars' atmosphere traveling 21,000 kilometers per hour (13,000 miles per hour). It began to slow, as friction from the atmosphere dragged against it. But friction also caused intense heating. How would *Curiosity* land safely?

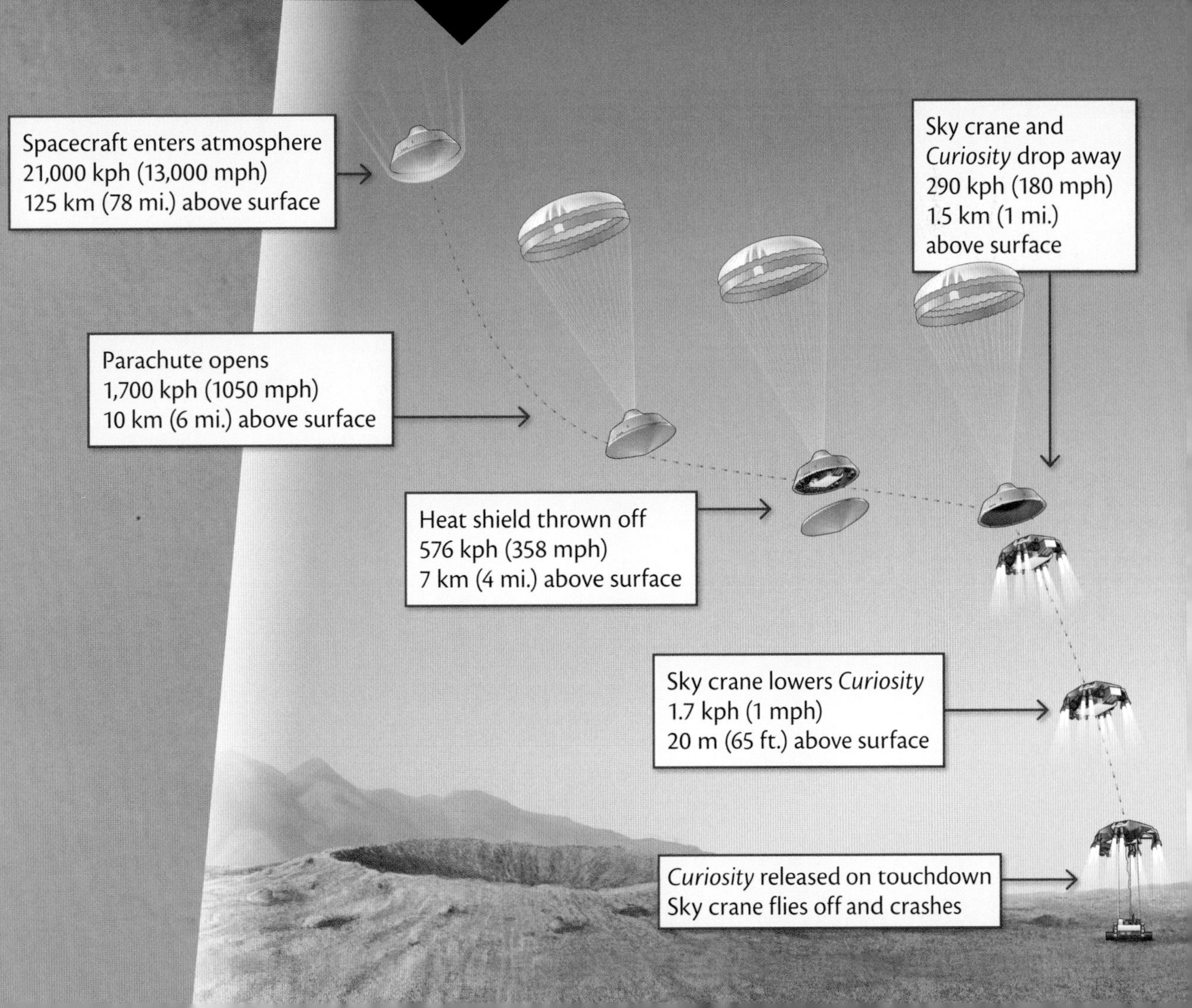

NASA scientists believe Gale Crater might have contained water in the past. It would be a good place to look for traces of water and carbon, a mineral needed for life. That's exactly what scientists had *Curiosity* start doing. After a few weeks of systems testing, scientists took the rover on its first drive. *Curiosity is* now on the move and scientists are directing it to collect and test samples from the soil and atmosphere.

Curiosity is the first rover able to test samples right on board. Ten scientific instruments and 17 cameras help the rover do its work. Take a look at this drawing of how *Curiosity* would look on Mars' surface. Then read on to compare this drawing with *Curiosity*'s self-portrait.

CURIOSITY: A CLOSER LOOK

LENGTH: 3 M (10 FT.)
ARM REACH: 2.2 M (7 FT.)
WEIGHT: 910 KG (2,000 LBS.)
WIDTH: 2.7 M (9 FT.)
HEIGHT: 2.2 M (7 FT.)
TOP SPEED: 4 CM (1.5 IN.) PER SECOND

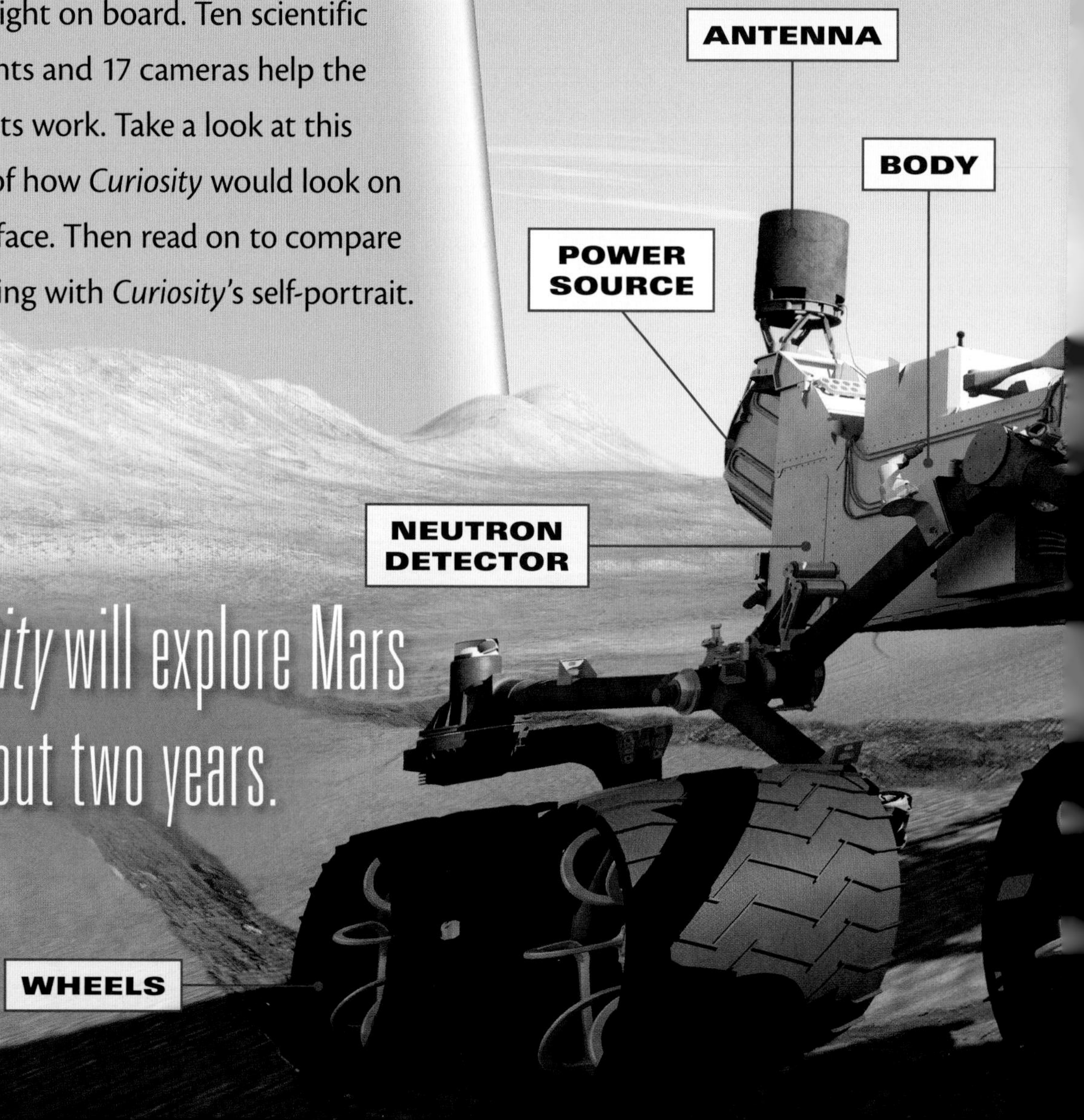

Curiosity will explore Mars for about two years.

SIX WHEELS on jointed "legs" allow *Curiosity* to climb slopes and roll over rocks as big as the wheels. By locking five of the wheels, *Curiosity* can use the sixth to dig below the surface.

A **ROBOTIC ARM** scoops up rocks and soil and delivers samples inside for testing. It also has a magnifier camera and X-ray equipment to study minerals in each sample.

The **MAST** uses six cameras as "eyes" to navigate.

Four pairs of **HAZCAMS** at the front and rear watch out for hazards.

The **CHEMCAM** uses a laser to burn tiny holes in rocks and soil a few meters ahead. The way a sample burns reveals the chemicals it contains.

An on-board **POWER SOURCE** changes heat into energy that is stored in batteries. The batteries then create **electric current** for the rover's systems.

A **WEATHER STATION** records weather conditions each hour and the amount of ultraviolet radiation from the sun.

An **OBSERVATION TRAY** gives two tools on the robotic arm a place to examine samples.

A **NEUTRON DETECTOR** checks for water in rocks and soil.

An **ANTENNA** allows for communication with scientists on Earth.

Inside the **BODY,** one computer and a backup computer monitor all of the systems in the rover.

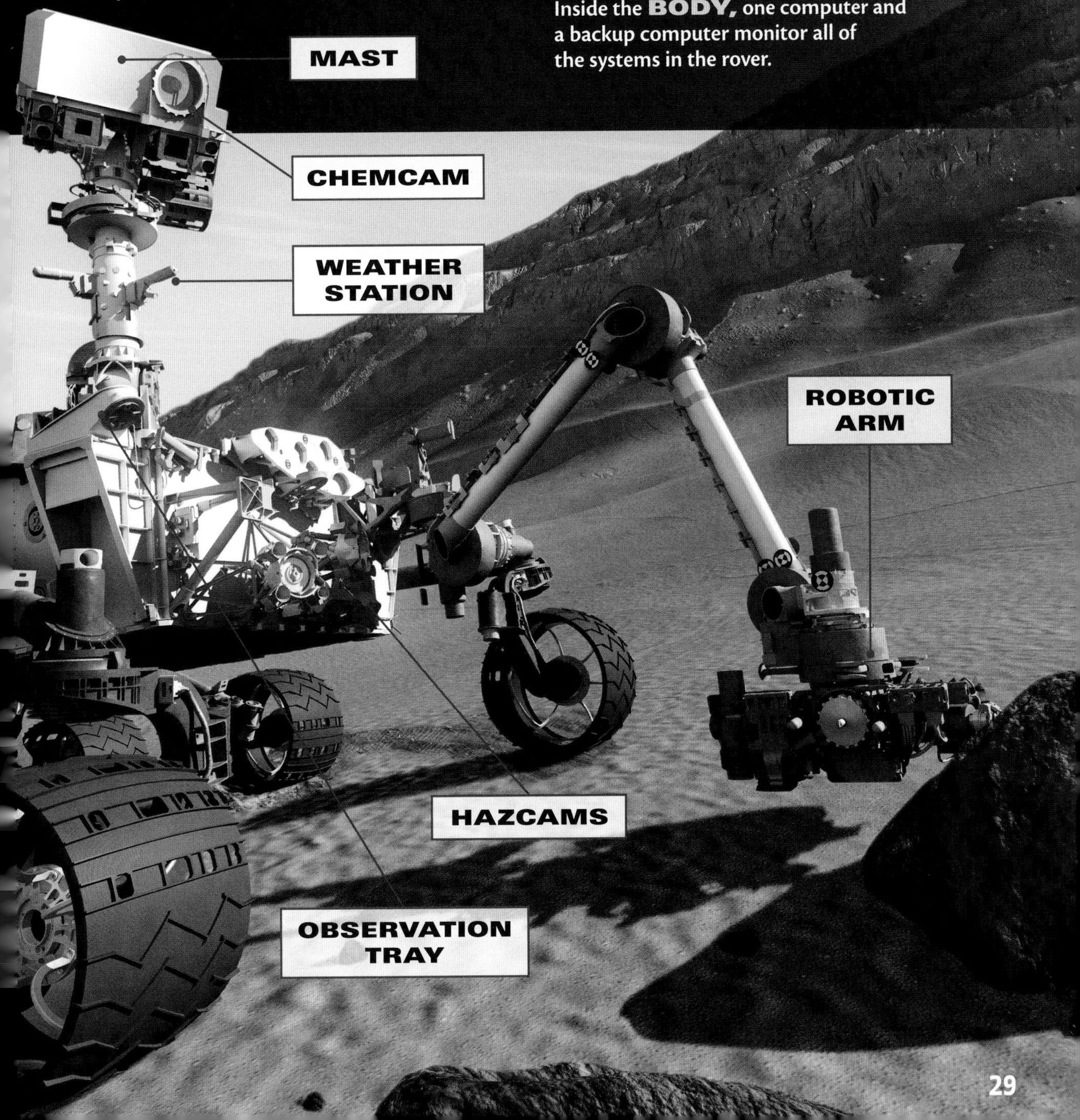

Curiosity sent more than 23,000 images back to Earth in its first 16 weeks.

Curiosity took this self-portrait on the 84th and 85th day of its mission. The camera on the end of the rover's robotic arm took more than 50 pictures from different positions. The pictures were then combined to create a single image.

EARLY DISCOVERIES

Curiosity's early tests ran smoothly. First, the rover spent a few weeks collecting sand samples from an area called Rocknest. Its robotic arm also blasted a rock with its laser beam so it could analyze the microscopic dust. *Curiosity* then started moving. It discovered an ancient streambed and monitored a huge, distant dust storm. Instruments inside the rover also further analyzed the Rocknest samples. They identified traces of water in the sand. When heated, the sand also formed a chemical containing a tiny amount of carbon as a side product. Carbon is a building block for life. However, scientists say it's too soon to know how this carbon got into the soil. The rover continues inching toward its main target, a five-kilometer- (three-mile-) high mountain named Mount Sharp. Scientists can hardly wait to see what awaits *Curiosity* there!

SLOW BUT SURE—123 DAYS, 598 M (1,961 FT.)

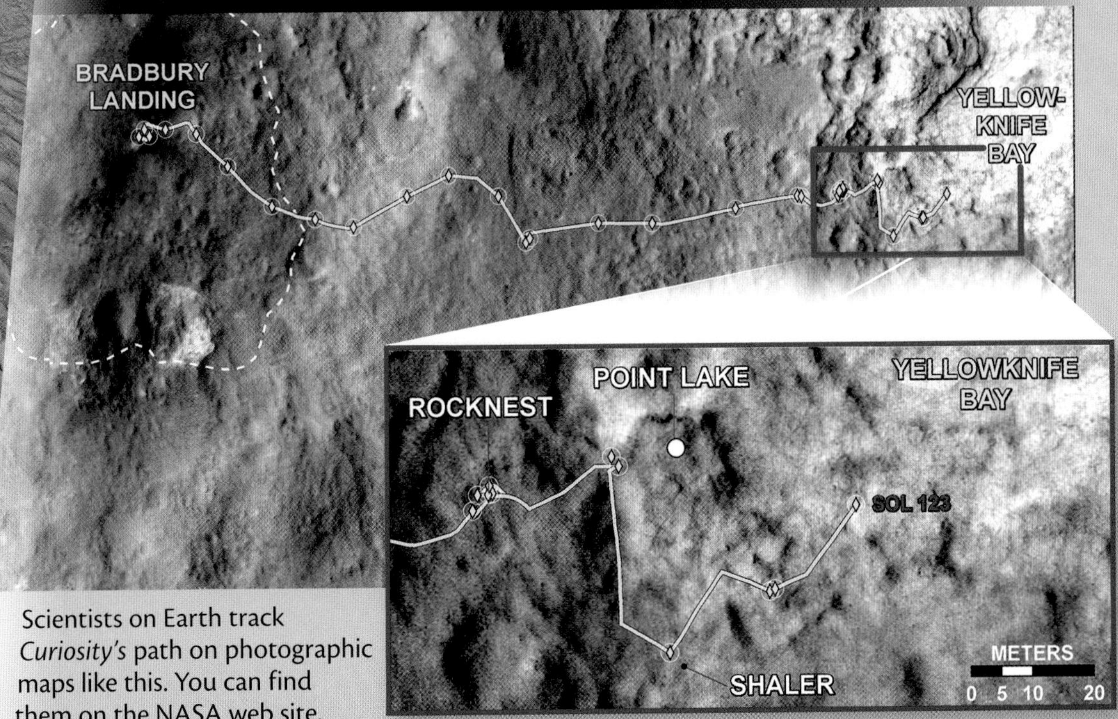

Scientists on Earth track *Curiosity*'s path on photographic maps like this. You can find them on the NASA web site.

Check In How was *Curiosity* engineered to travel across Mars' surface?

Discuss

1. What connections can you make among the four pieces in *Let's Keep Moving*?

2. What surprised you about the design and technology of the early cars?

3. Which piece contained the most complex problem? How did the scientists attempt to solve it? Why was it so difficult?

4. What have we learned from *Curiosity* so far? What more do scientists hope to learn?

5. What do you still wonder about ways people get from one place to another? What would be some good ways to find more information?